Math
ADVANTAGE

Teaching Resources

Grade 3

Harcourt Brace & Company

Orlando • Atlanta • Austin • Boston • San Francisco • Chicago • Dallas • New York • Toronto • London

http://www.hbschool.com

CONTENTS

TIME, MONEY, MEASUREMENT

DATA, PROBABILITY, AND STATISTICS

For Use with Specific Activities

PUPIL'S EDITION LESSONS

TEACHER'S EDITION PRACTICE GAMES

TEACHER'S EDITION, TAB B, PRACTICE ACTIVITIES

LEARNING CENTER CARDS

PROBLEM-SOLVING
THINK ALONG

Problem Solving

Understand

1. Retell the problem in your own words. _____

_____

_____

_____

2. List the information given. _____

_____

3. Restate the question as a fill-in-the-blank sentence. _____

_____

Plan

4. List one or more problem-solving strategies that you can use. _____

_____

5. Predict what your answer will be. _____

_____

Solve

6. Show how you solved the problem. _____

_____

_____

_____

7. Write your answer in a complete sentence. _____

_____

Look Back

8. Tell how you know your answer is reasonable. _____

_____

9. Describe another way you could have solved the problem. _____

_____

PROBLEM-SOLVING
THINK ALONG

Problem-Solving Think Along

Understand

    1. What is the problem about?

    2. What information is given in the problem?

    3. What is the question?

Plan

    4. What problem-solving strategies might I try to help me solve the problem?

    5. About what do I think my answer will be?

Solve

    6. How can I solve the problem?

    7. How can I state my answer in a complete sentence?

Look Back

    8. How do I know whether my answer is reasonable?

    9. How else might I have solved this problem?

4	0
5	1
6	2
7	3

12	8
13	9
14	10
15	11

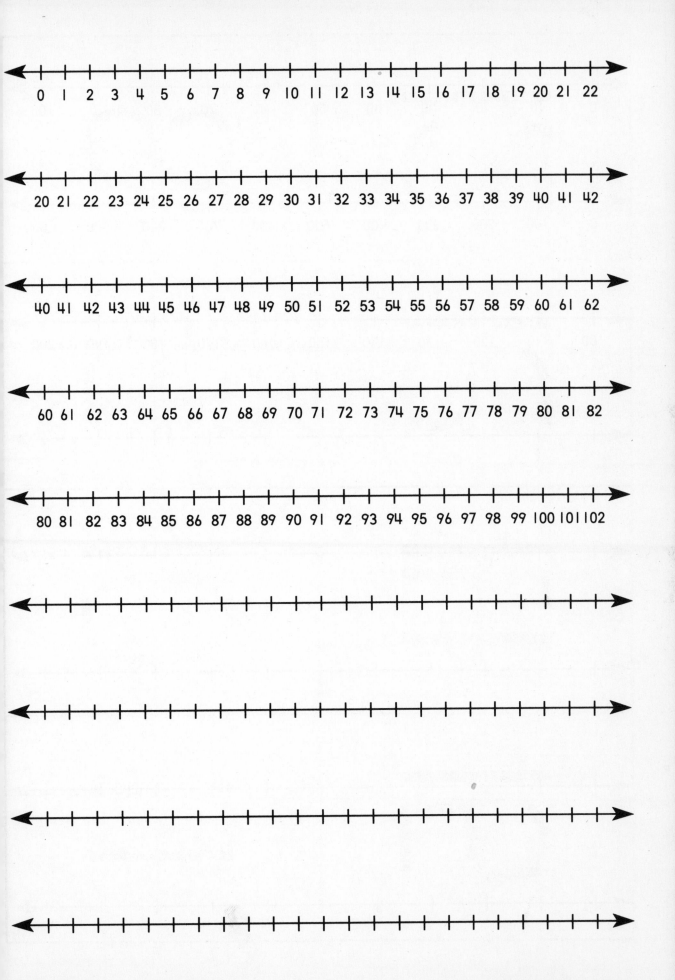

0 1 2 3 4 5 6 7 8 9 10 11 12 13 14 15 16 17 18 19 20 21 22

20 21 22 23 24 25 26 27 28 29 30 31 32 33 34 35 36 37 38 39 40 41 42

40 41 42 43 44 45 46 47 48 49 50 51 52 53 54 55 56 57 58 59 60 61 62

60 61 62 63 64 65 66 67 68 69 70 71 72 73 74 75 76 77 78 79 80 81 82

80 81 82 83 84 85 86 87 88 89 90 91 92 93 94 95 96 97 98 99 100 101 102

Number Lines

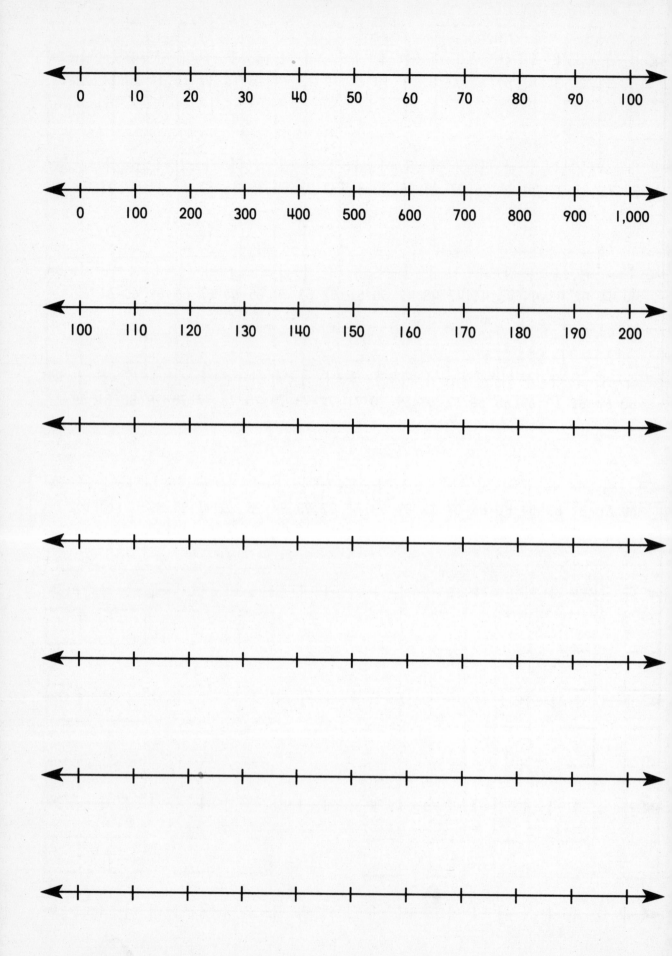

0   10   20   30   40   50   60   70   80   90   100

0   100   200   300   400   500   600   700   800   900   1,000

100   110   120   130   140   150   160   170   180   190   200

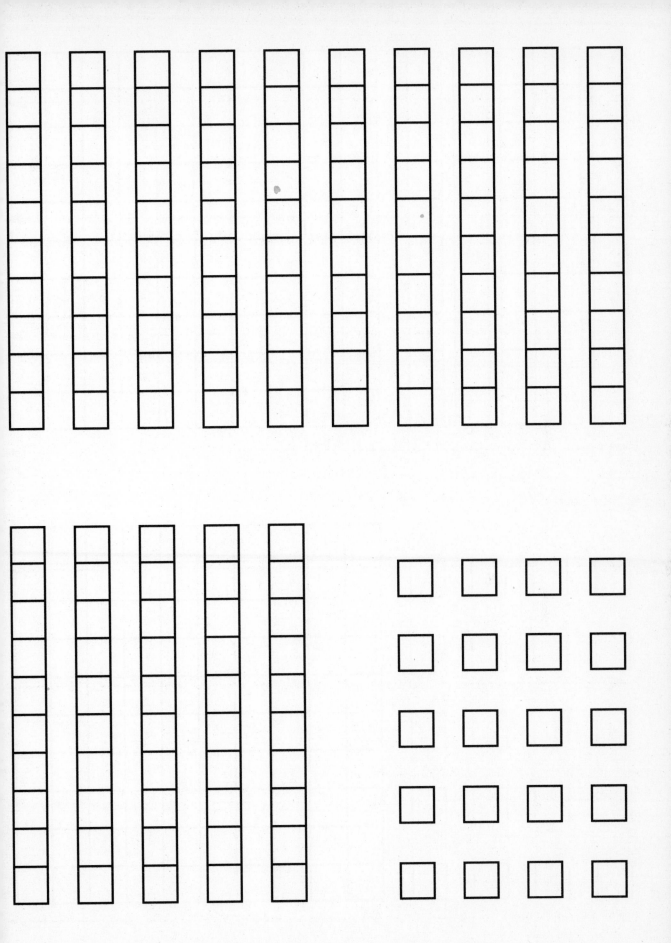

Base-Ten Materials

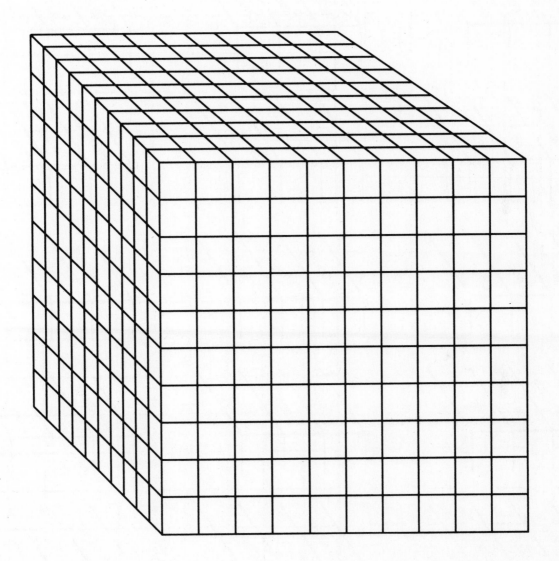

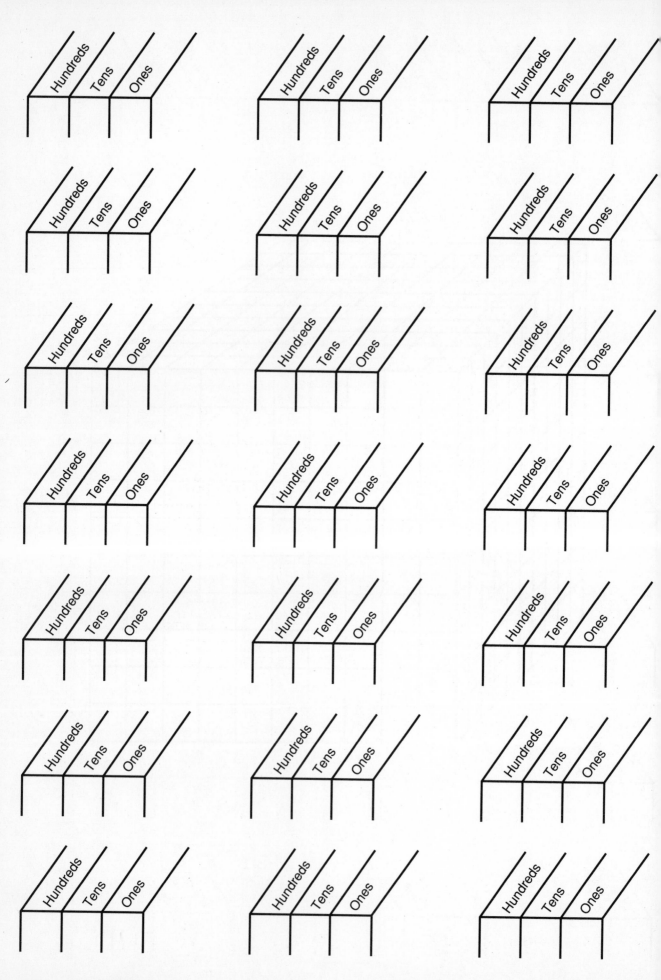

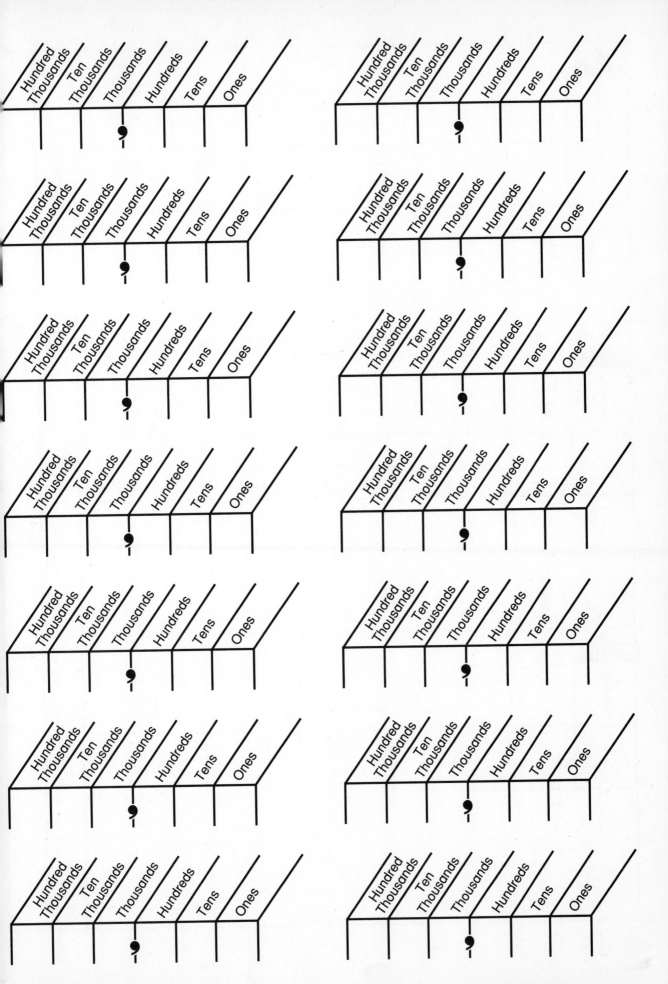

Period:

Hundreds

Regroup

Tens

Regroup

Ones

Regroup

hundreds	tens	ones

_____

hundreds	tens	ones

_____

hundreds	tens	ones

_____

hundreds	tens	ones

_____

hundreds	tens	ones

_____

hundreds	tens	ones

_____

hundreds	tens	ones

_____

hundreds	tens	ones

_____

Workmat 5

Hundreds	Tens	Ones

1	2	3	4	5	6	7	8	9	10
11	12	13	14	15	16	17	18	19	20
21	22	23	24	25	26	27	28	29	30
31	32	33	34	35	36	37	38	39	40
41	42	43	44	45	46	47	48	49	50
51	52	53	54	55	56	57	58	59	60
61	62	63	64	65	66	67	68	69	70
71	72	73	74	75	76	77	78	79	80
81	82	83	84	85	86	87	88	89	90
91	92	93	94	95	96	97	98	99	100

Hundreds Chart

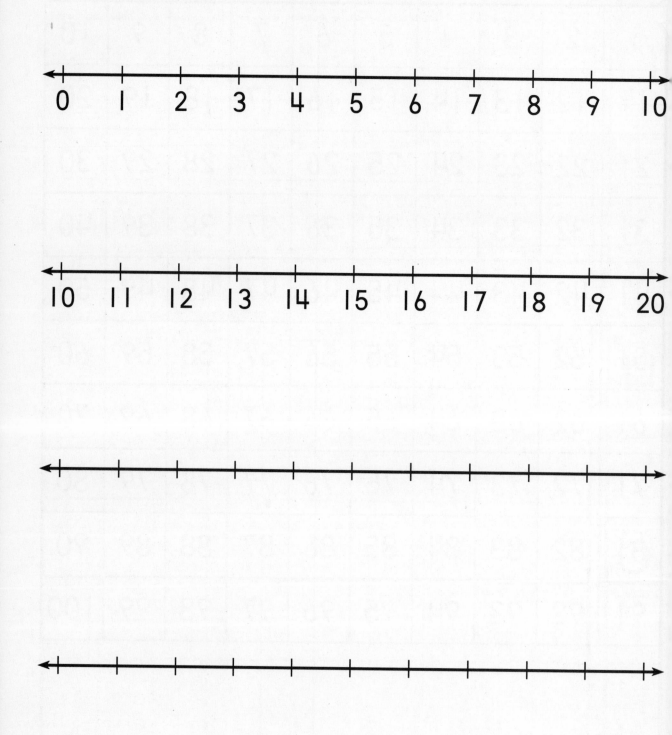

	0	1	2	3	4	5	6	7	8	9
0										
1										
2										
3										
4										
5										
6										
7										
8										
9										

Addition/Multiplication Table

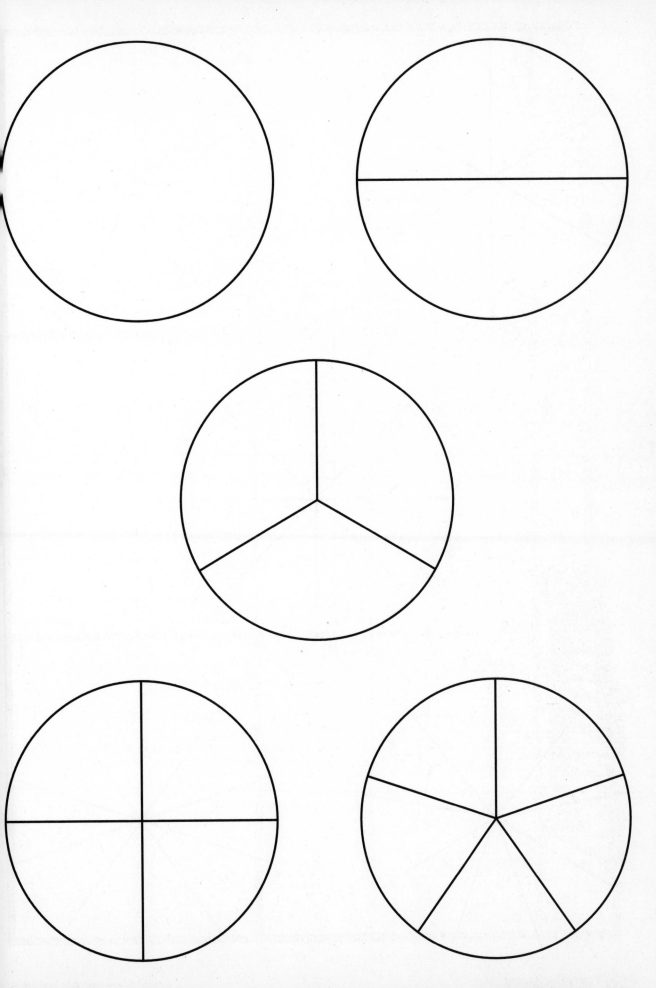

Fraction Circles

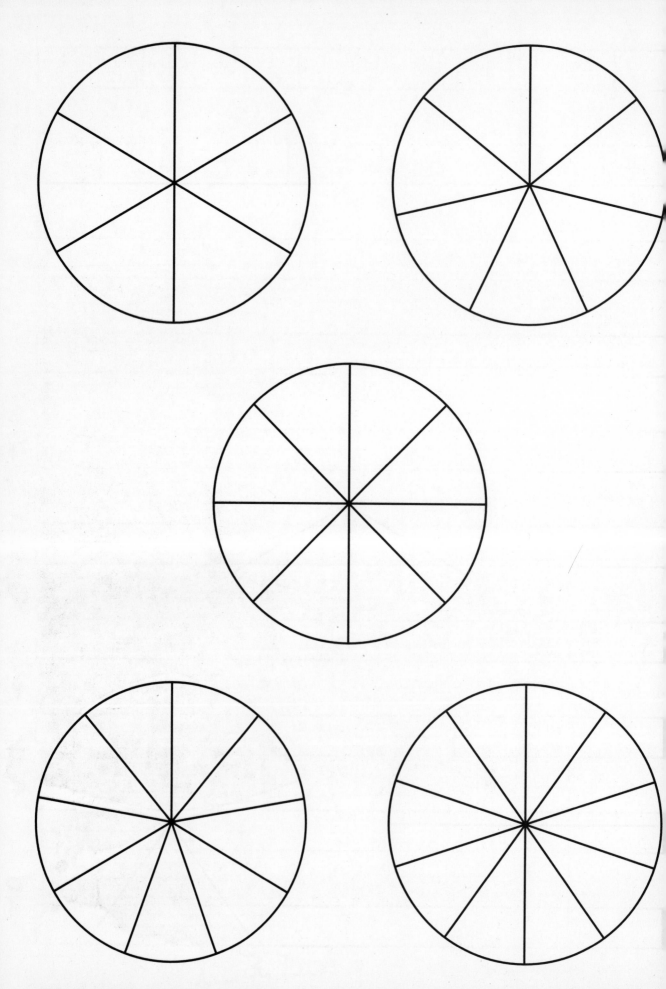

Fraction Circles

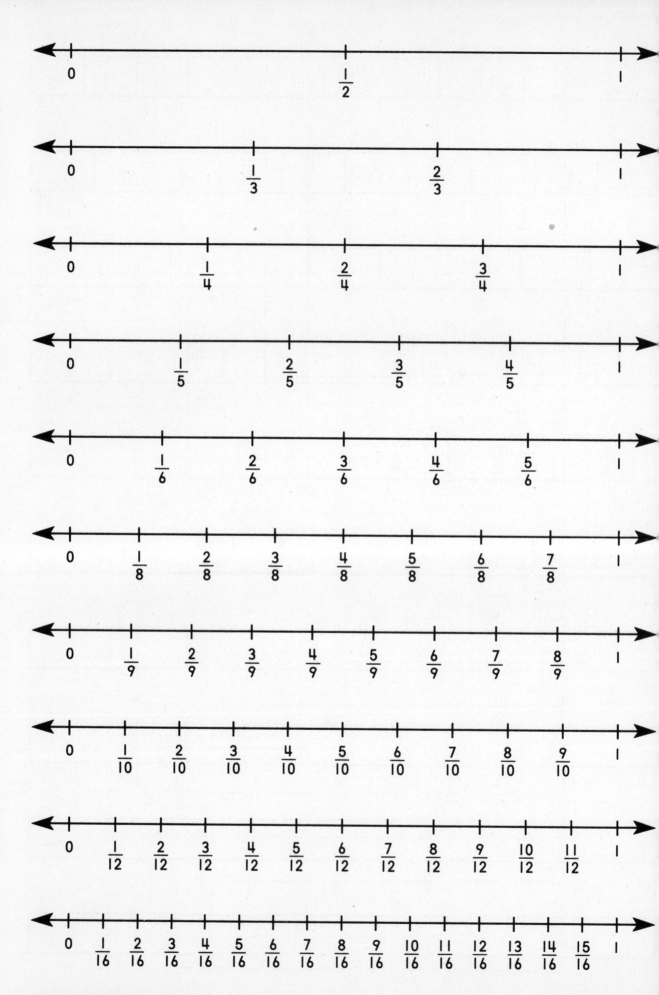

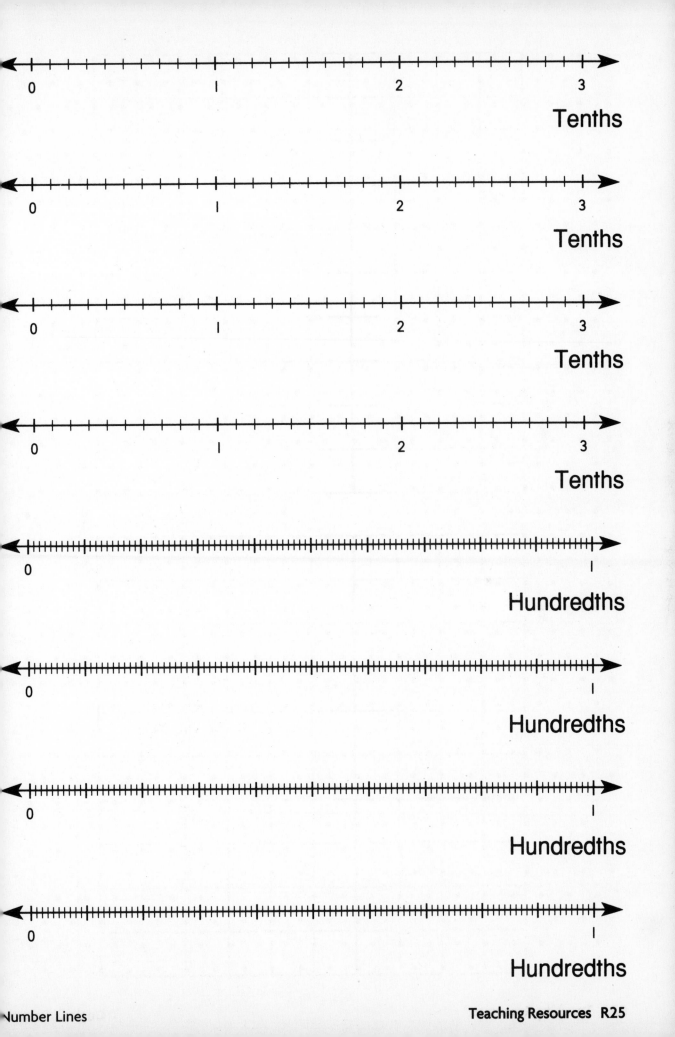

Tenths

Tenths

Tenths

Tenths

Hundredths

Hundredths

Hundredths

Hundredths

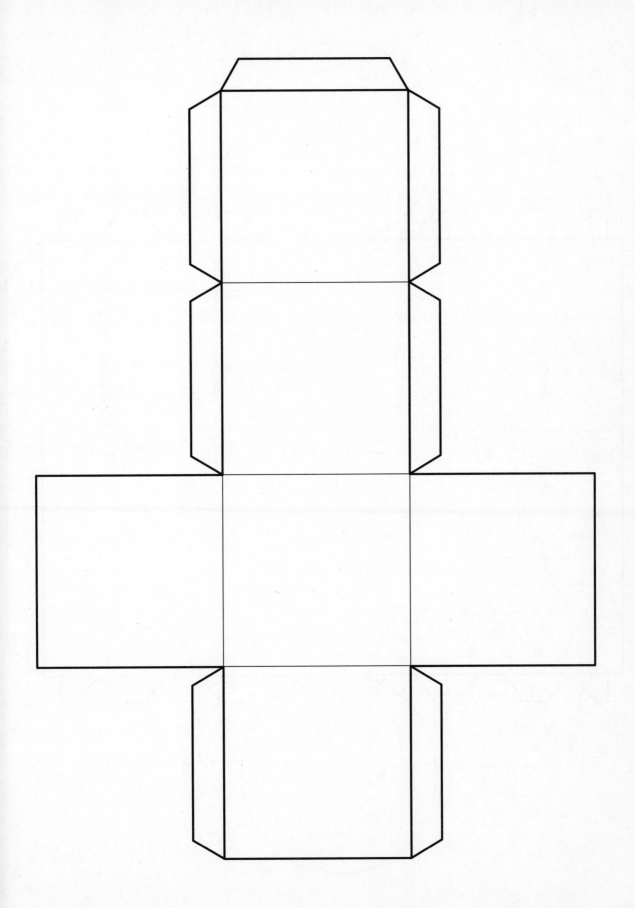

Cube Pattern

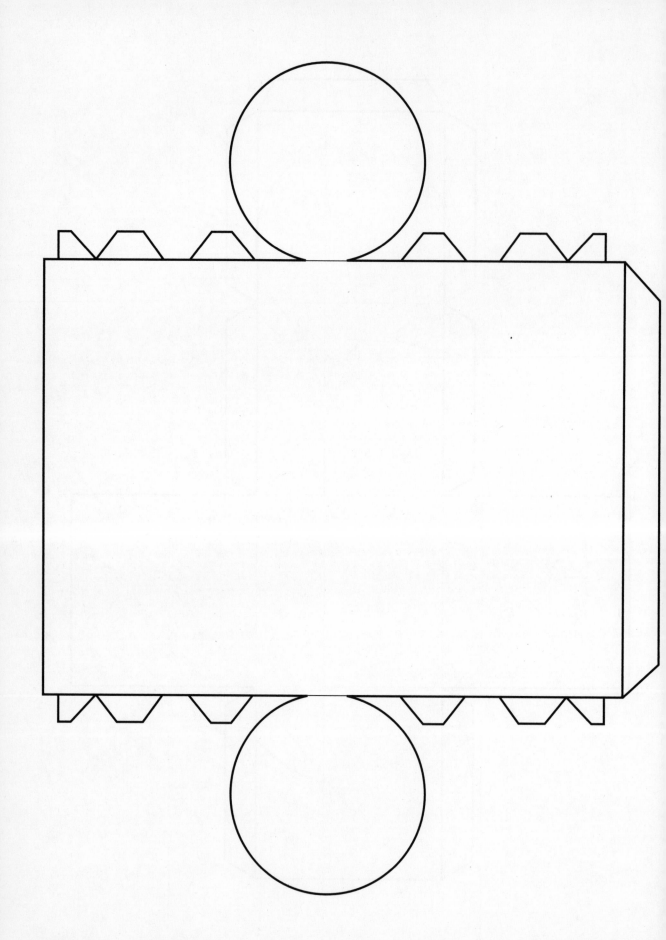

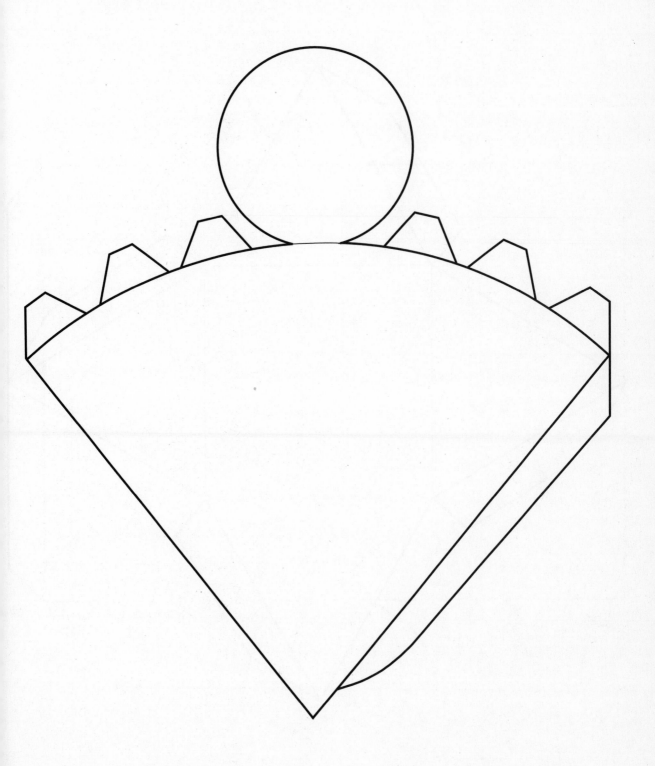

Cone Pattern

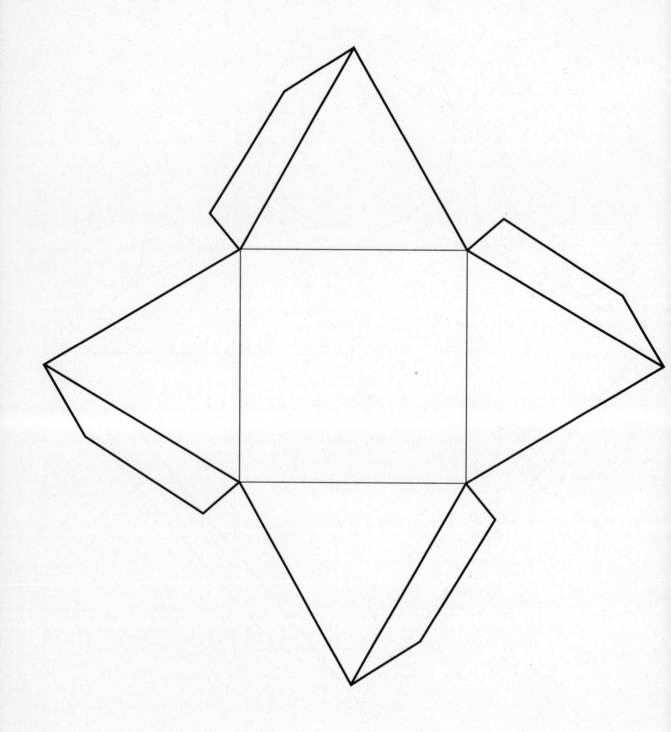

Square Pyramid Pattern

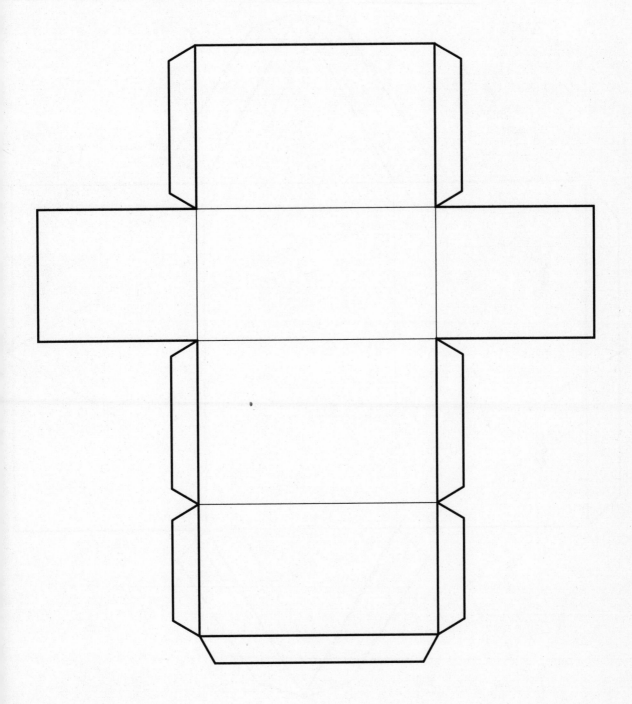

Rectangular Prism Pattern

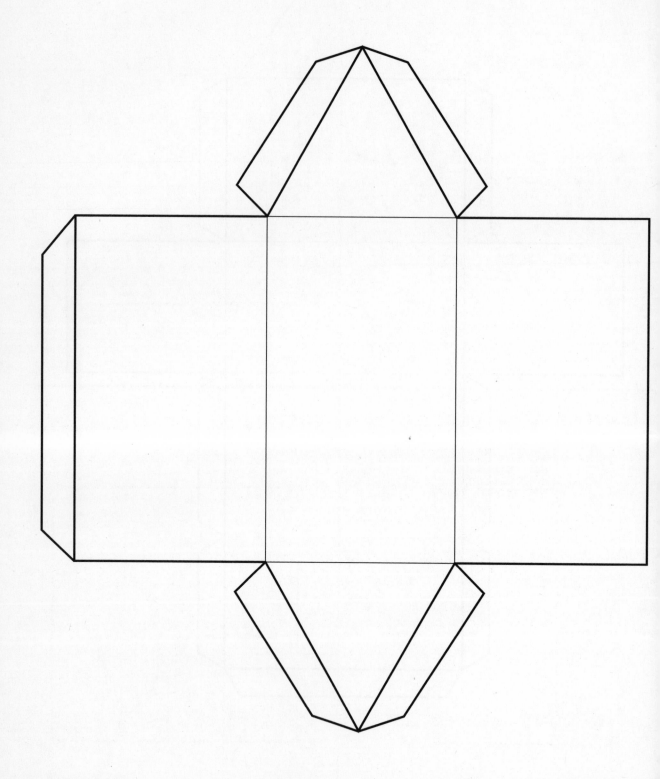

Triangular Prism Pattern

Squares and Rectangles

Square Dot Paper

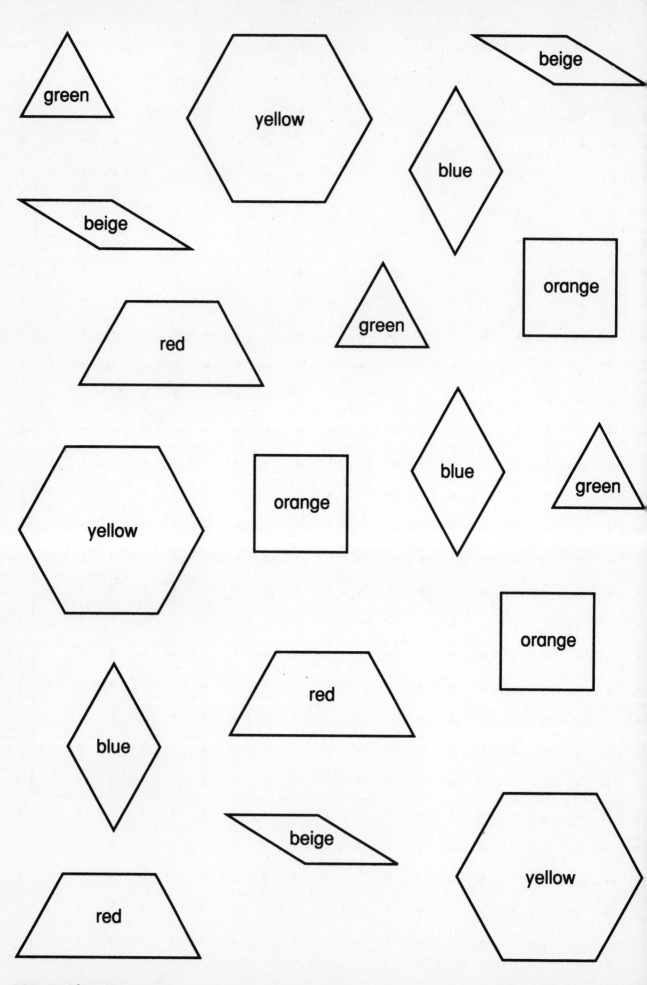

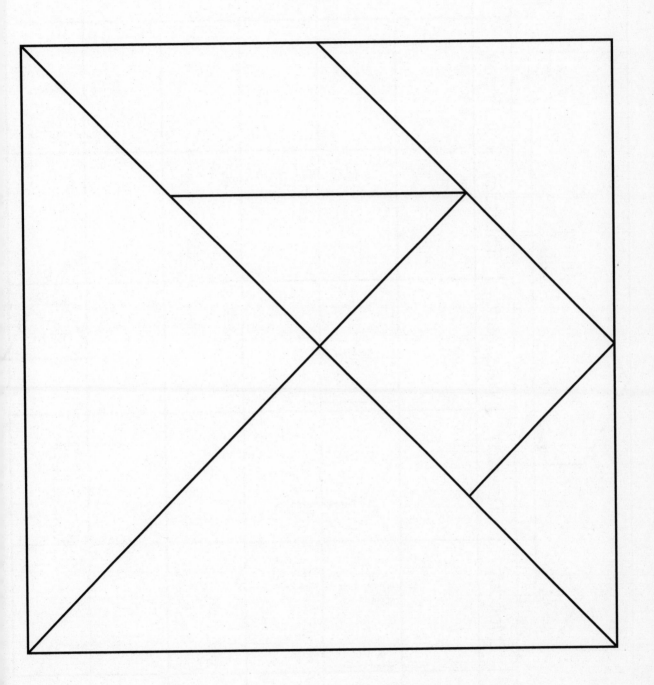

Tangram Pattern

Sunday	Monday	Tuesday	Wednesday	Thursday	Friday	Saturday

Blank Calendar

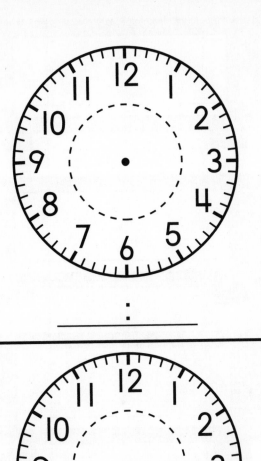

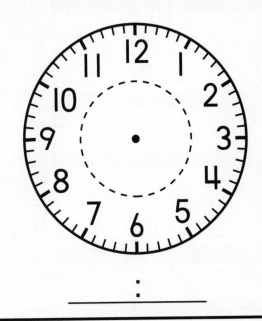

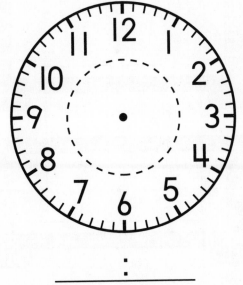

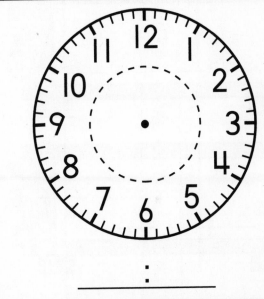

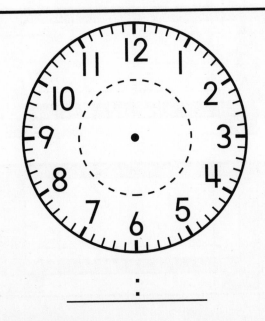

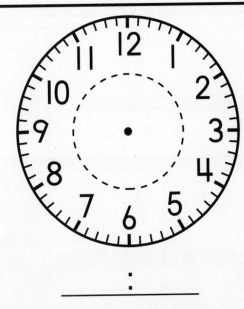

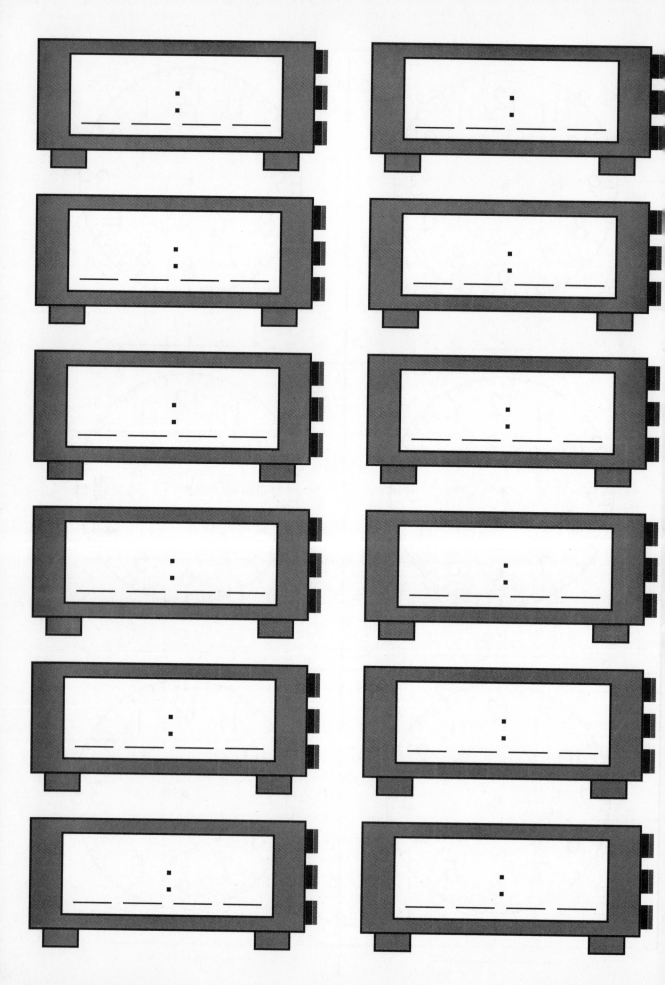

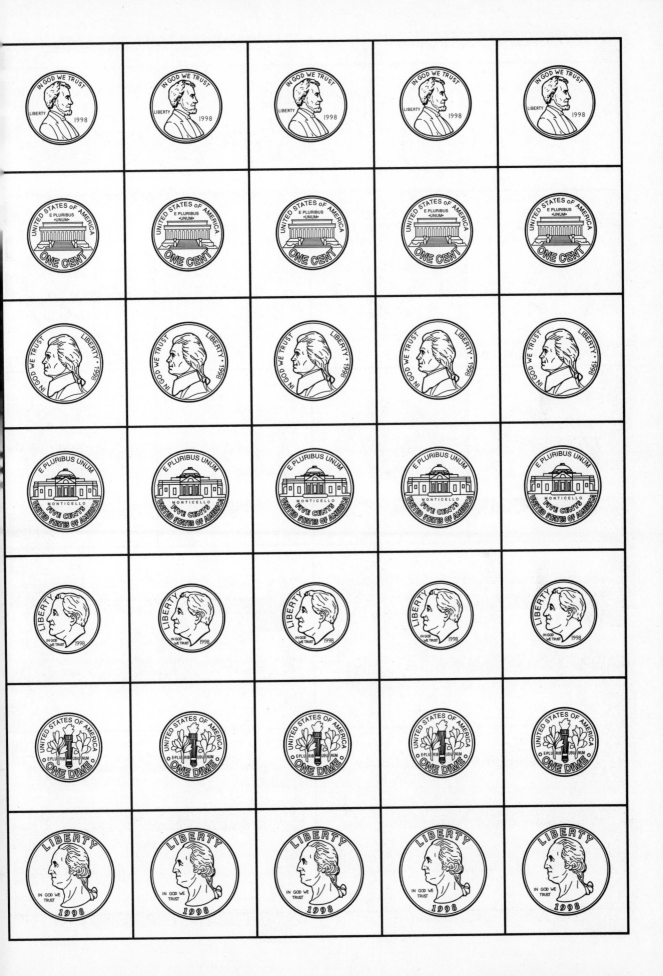

Coins

Coins and Bills

Bills

Workmat 4

Quarter	Dime	Nickel	Penny

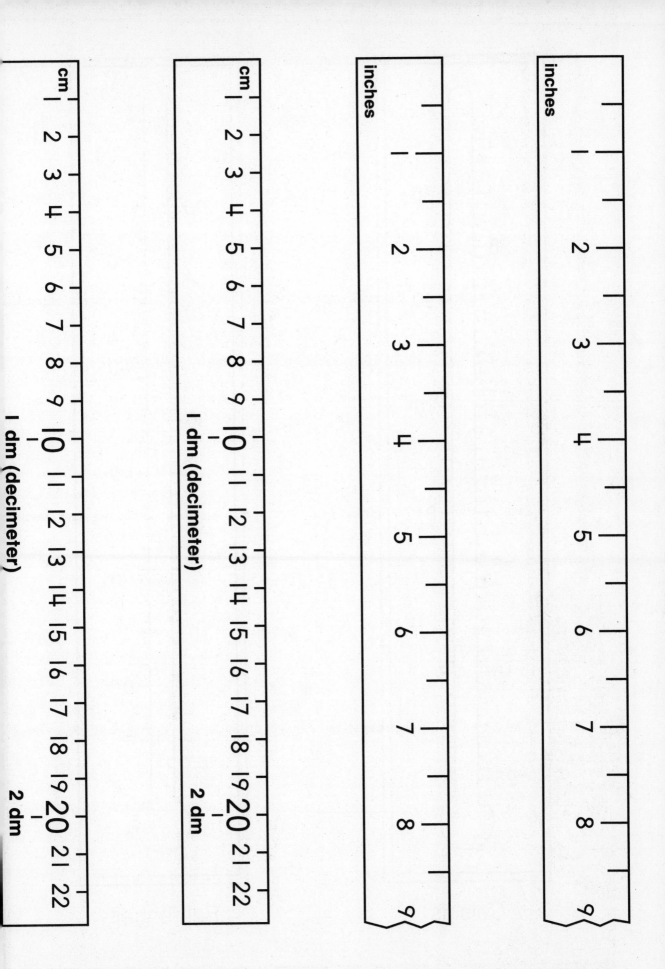

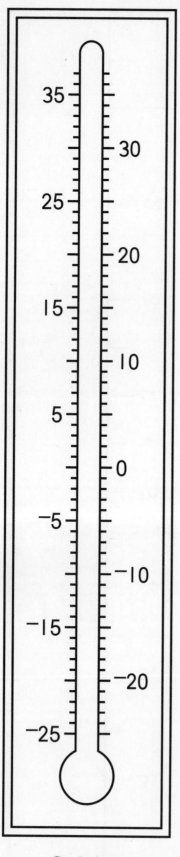

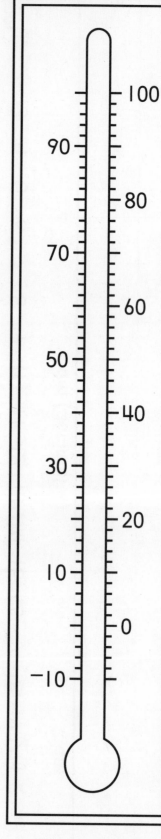

Celsius

_____ °C

Fahrenheit

_____ °F

Title: _____

=

Title: _____

Scale Label: _____

Date Label: _____

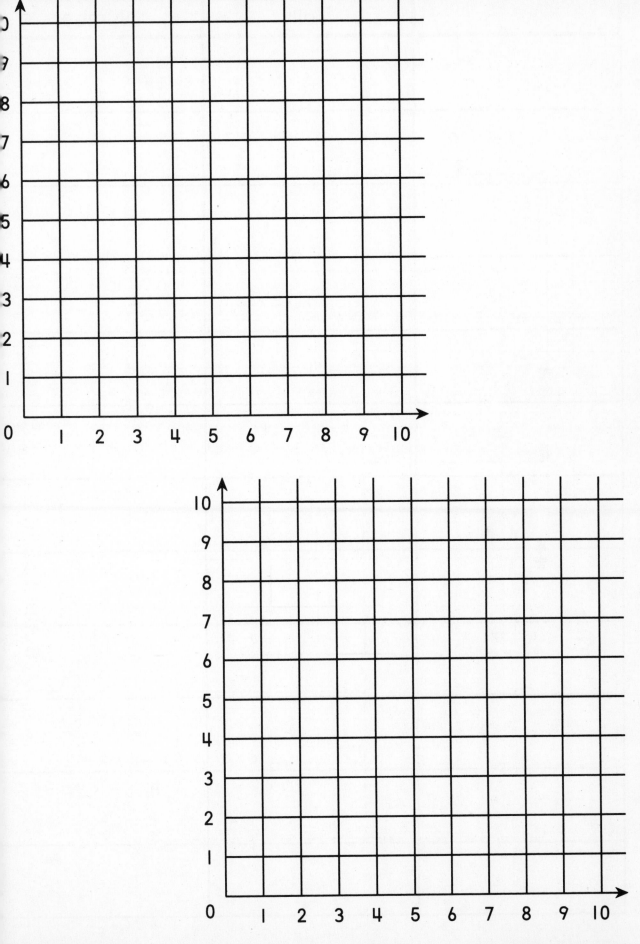

Grid of Quadrant 1

	Tally	Number

Tally Table

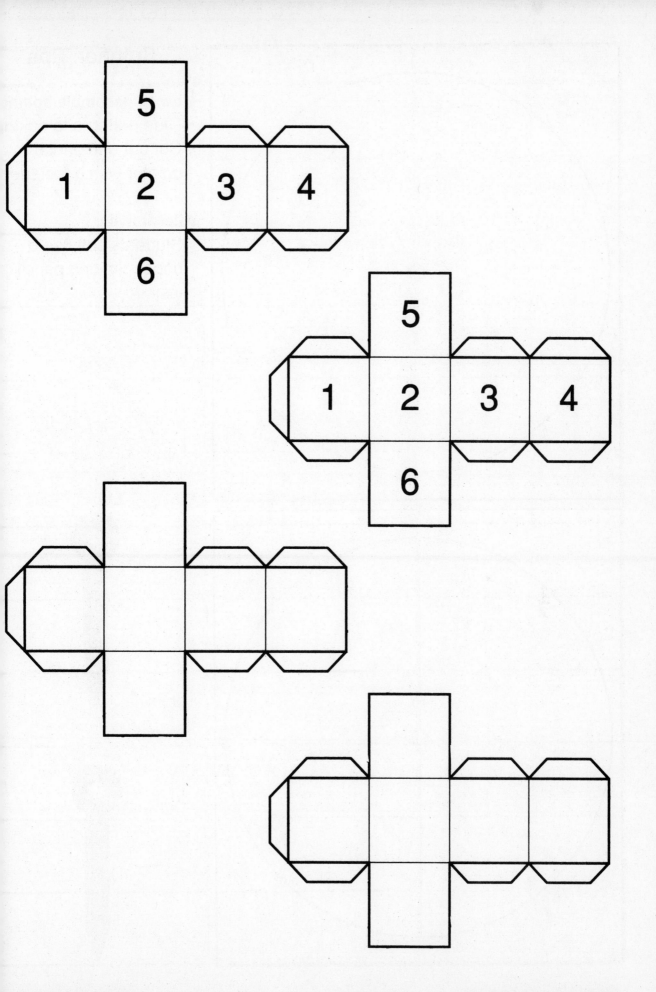

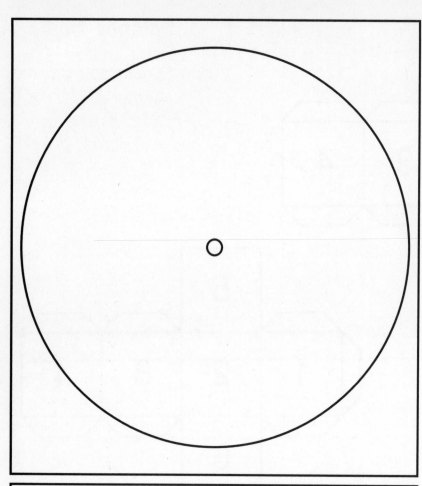

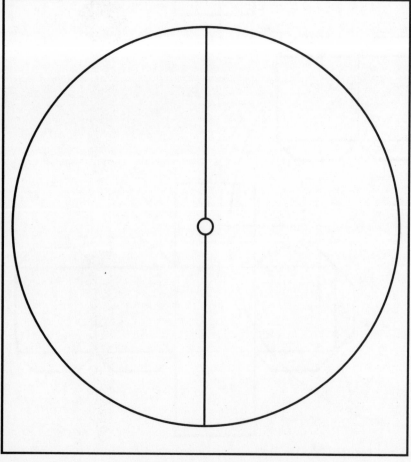

Spinner Tips

How to assemble spinne

- Glue patterns to oakta
- Cut out and attach pointer with a fastener

Alternative

- Students can use a paper clip and pencil instead.

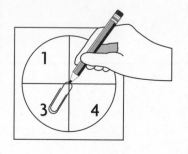

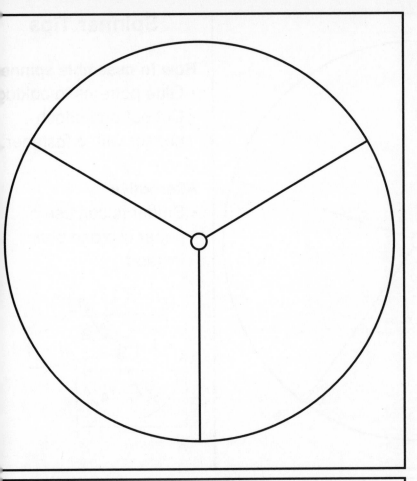

Spinner Tips

How to assemble spinner.
- Glue patterns to oaktag.
- Cut out and attach pointer with a fastener.

Alternative
- Students can use a paper clip and pencil instead.

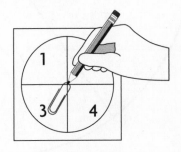

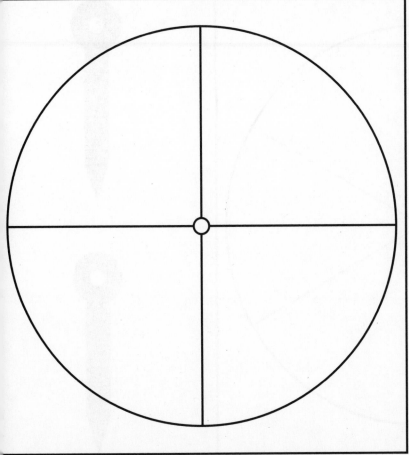

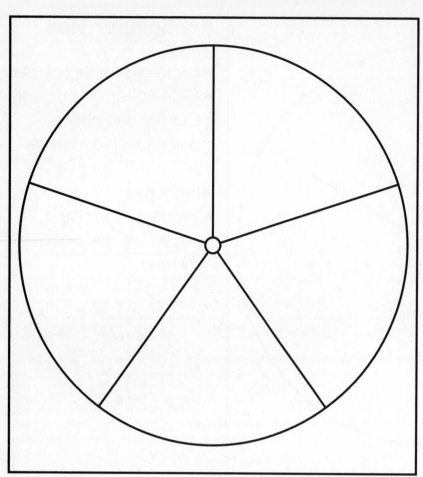

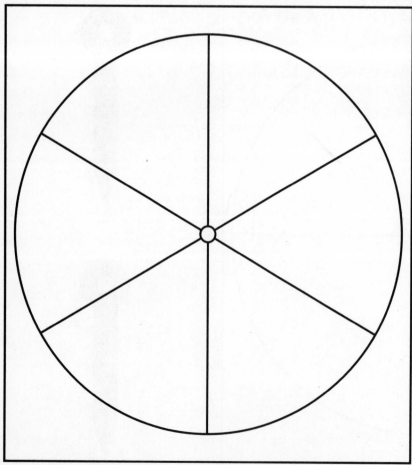

Spinner Tips

How to assemble spinner

- Glue patterns to oaktag.
- Cut out and attach pointer with a fastener.

Alternative

- Students can use a paper clip and pencil instead.

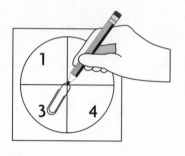

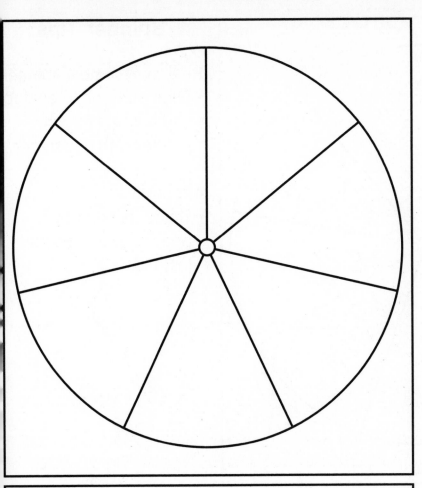

Spinner Tips

How to assemble spinner.
- Glue patterns to oaktag.
- Cut out and attach pointer with a fastener.

Alternative
- Students can use a paper clip and pencil instead.

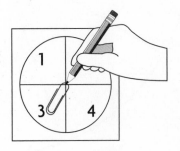

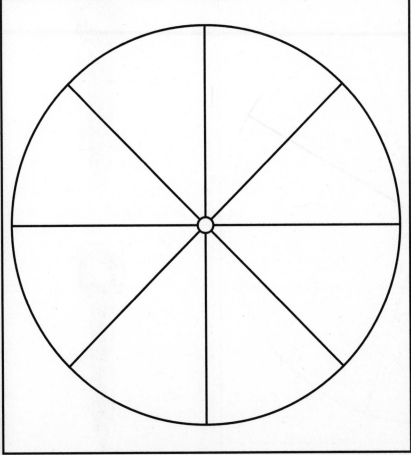

Spinners (7- and 8-section)

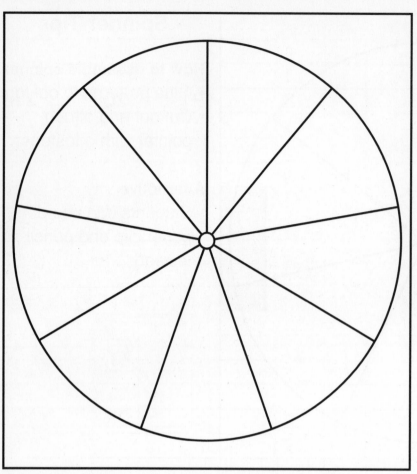

Spinner Tips

How to assemble spinner
- Glue patterns to oaktag.
- Cut out and attach pointer with a fastener.

Alternative
- Students can use a paper clip and pencil instead.

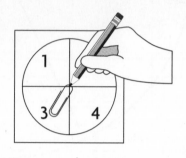

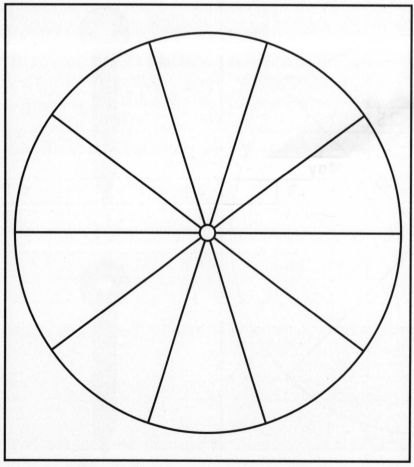

January

1	2	3	4	5	6	7
8	9	10	11	12	13	14
15	16	17	18	19	20	21
22	23	24	25	26	27	28
29	30	31				

February

			1	2	3	4
5	6	7	8	9	10	11
12	13	14	15	16	17	18
19	20	21	22	23	24	25
26	27	28				

March

			1	2	3	4
5	6	7	8	9	10	11
12	13	14	15	16	17	18
19	20	21	22	23	24	25
26	27	28	29	30	31	

April

						1
2	3	4	5	6	7	8
9	10	11	12	13	14	15
16	17	18	19	20	21	22
23	24	25	26	27	28	29
30						

May

	1	2	3	4	5	6
7	8	9	10	11	12	13
14	15	16	17	18	19	20
21	22	23	24	25	26	27
28	29	30	31			

June

				1	2	3
4	5	6	7	8	9	10
11	12	13	14	15	16	17
18	19	20	21	22	23	24
25	26	27	28	29	30	

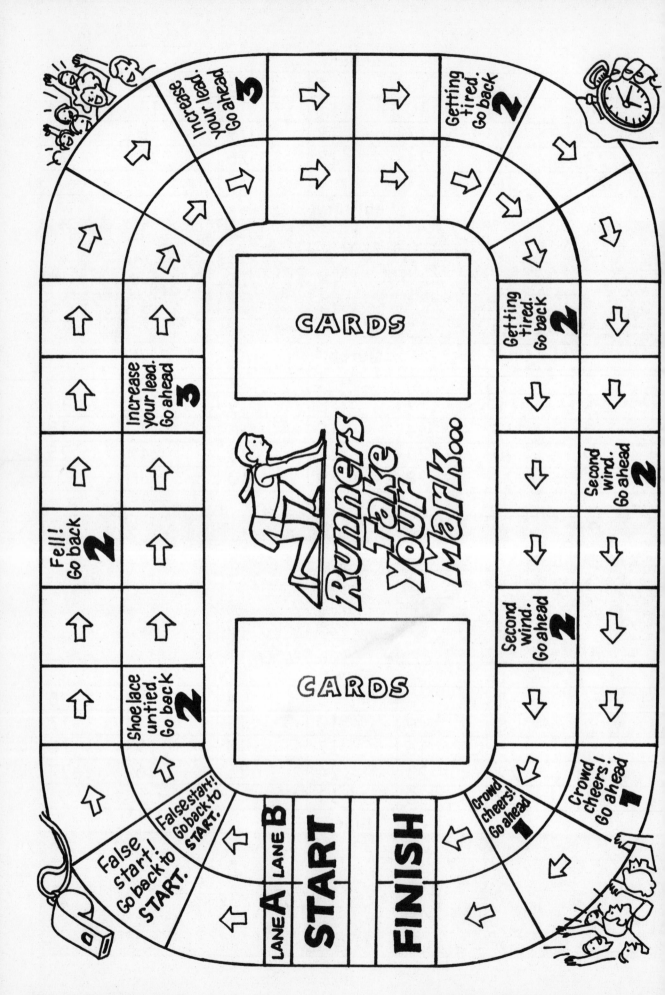

9	11	17	19	22	26
32	35	40	41	45	48
53	57	60	63	64	68
70	72	75	77	78	80
81	84	85	88	90	96

Activity Card 1

1. Say the twelve months of the year in order, three times.

Activity Card 2

2. Use the movie schedule to answer both questions.

Movie Schedule	Day	Time
The Big Cat	Fri	4:00–6:00
The Red Tomato	Sat	4:00–5:45
Lizard	Sat	5:30–7:15
My Old Shoe	Sun	4:00–5:30

Aunt Margaret will take you to the movies at 4:00 P.M. on Saturday. Which movie will you see? How long does the movie last?

Activity Card 3

3. Use the schedule to answer the questions.

Monday's Schedule	
Activity	**Time**
Morning announcements	8:15–8:30
Reading	8:30–9:15
Music	9:15–10:00
Language arts	10:00–11:00
Science	11:00–11:30
Lunch	11:30–12:15

How long is Science class? Which activities are 45 minutes long?

Activity Card 4

4. Use the calendar to answer the questions.

January						
Sun	Mon	Tue	Wed	Thur	Fri	Sat
			1	2	3	4
5	6	7	8	9	10	11
12	13	14	15	16	17	18
19	20	21	22	23	24	25
26	27	28	29	30	31	

On what day of the week is February 1? On what day of the week is New Year's Day? How many Saturdays are there in January?

Answer Card

1. January, February, March, April, May, June, July, August, September, October, November, December

2. *The Red Tomato*; 1 hour 45 minutes

3. 30 minutes; reading, music, lunch

4. Feb. 1 is Saturday; New Year's is Wednesday; 4 Saturdays

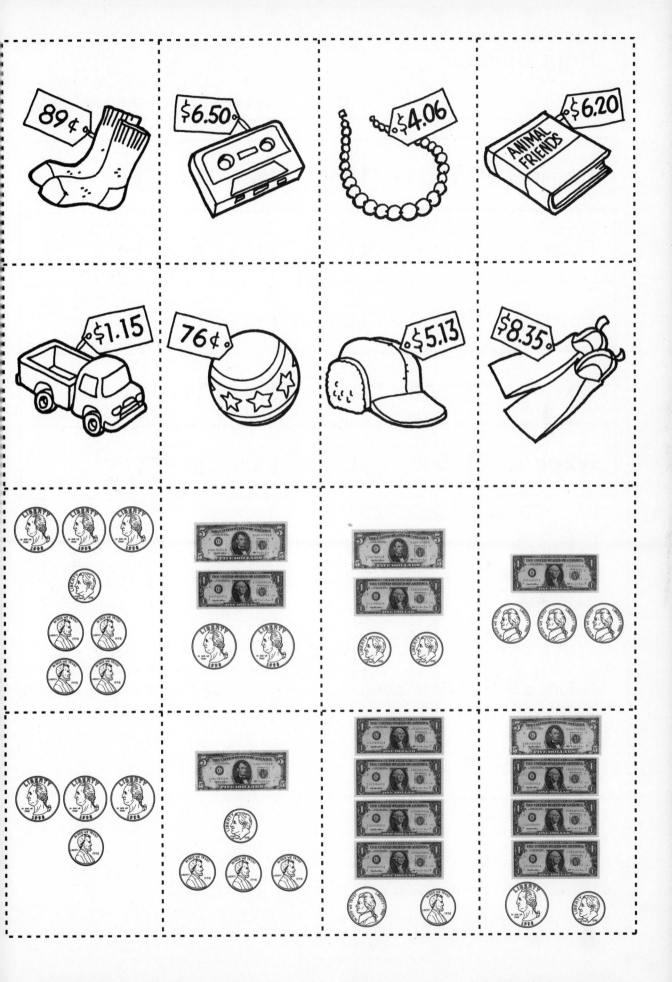

High Number

Name _____

Round 1	☐0,000 + ☐,000 + ☐00 + ☐0 + ☐ =
Round 2	☐0,000 + ☐,000 + ☐00 + ☐0 + ☐ =
Round 3	☐0,000 + ☐,000 + ☐00 + ☐0 + ☐ =
Round 4	☐0,000 + ☐,000 + ☐00 + ☐0 + ☐ =
Round 5	☐0,000 + ☐,000 + ☐00 + ☐0 + ☐ =

For use with Chapter 9 Practice Game

Walking the Line

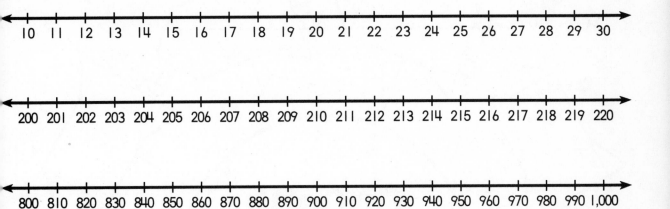

Which is the greatest number? 27, 22, 26	Which is the greatest number? 13, 23, 21	Round 27 to the nearest 10.
Round 12 to the nearest 10.	Round 16 to the nearest 10.	Which is the greatest number? 208, 204, 218
Which is the greatest number? 211, 214, 217	Round 208 to the nearest 10.	Round 217 to the nearest 10.
Round 219 to the nearest 100.	Round 878 to the nearest 100.	Round 849 to the nearest 10.
Round 959 to the nearest 100.	Round 975 to the nearest 100.	Which is the greatest number? 860, 960, 930

$45 \div 5$	$45 \div 9$	9×5	5×9
$35 \div 5$	$35 \div 7$	7×5	5×7
$28 \div 4$	$28 \div 7$	7×4	4×7
$12 \div 4$	$12 \div 3$	3×4	4×3
$27 \div 3$	$27 \div 9$	3×9	9×3
$20 \div 4$	$20 \div 5$	4×5	5×4
$18 \div 3$	$18 \div 6$	6×3	3×6
$6 \div 2$	$6 \div 3$	2×3	3×2
$10 \div 2$	$10 \div 5$	5×2	2×5
$8 \div 4$	$8 \div 2$	4×2	2×4

For use with Chapter 13 Practice Game

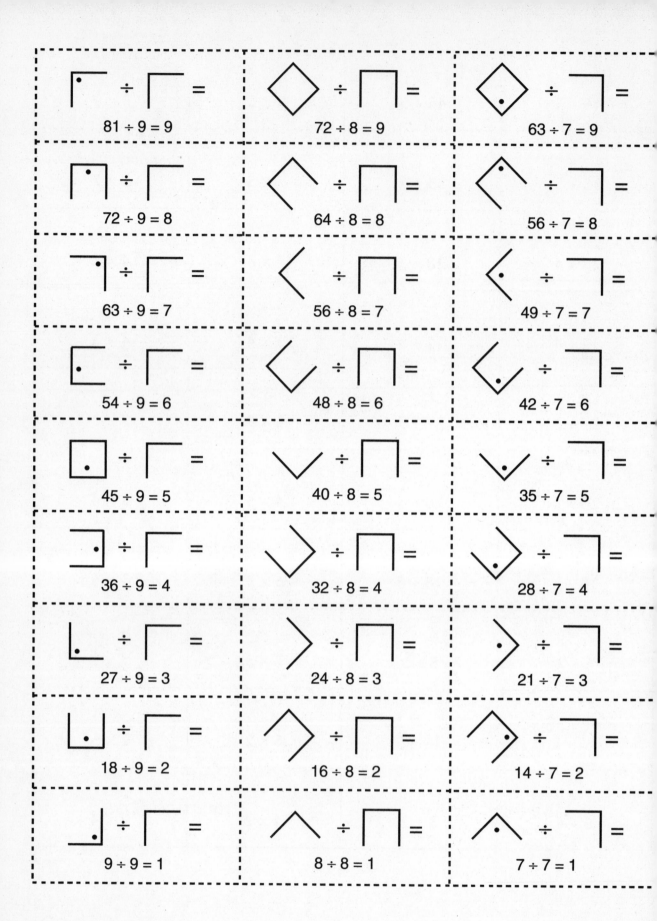

81 ÷ 9 = 9   72 ÷ 8 = 9   63 ÷ 7 = 9

72 ÷ 9 = 8   64 ÷ 8 = 8   56 ÷ 7 = 8

63 ÷ 9 = 7   56 ÷ 8 = 7   49 ÷ 7 = 7

54 ÷ 9 = 6   48 ÷ 8 = 6   42 ÷ 7 = 6

45 ÷ 9 = 5   40 ÷ 8 = 5   35 ÷ 7 = 5

36 ÷ 9 = 4   32 ÷ 8 = 4   28 ÷ 7 = 4

27 ÷ 9 = 3   24 ÷ 8 = 3   21 ÷ 7 = 3

18 ÷ 9 = 2   16 ÷ 8 = 2   14 ÷ 7 = 2

9 ÷ 9 = 1   8 ÷ 8 = 1   7 ÷ 7 = 1

1	2	3
4	5	6
7	8	9

Divisor Grid

9	18	27
36	45	54
63	72	81

Dividend Grid

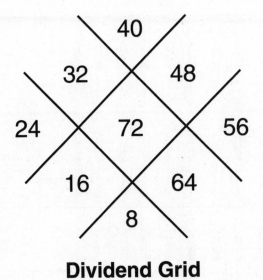

Dividend Grid

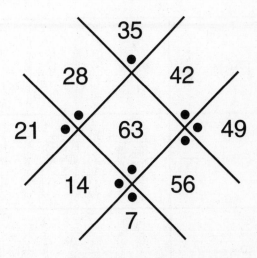

Dividend Grid

Foods We Eat

Fish	🧍🧍🧍🧍🧍🧍🧍🧍🧍
Chicken	🧍🧍🧍🧍🧍🧍🧍
Pasta	🧍🧍🧍🧍🧍
P.B. & J.	🧍🧍🧍🧍🧍🧍

Key: Each 🧍 equals 5 students

Foods We Eat

Fish	🧍🧍🧍🧍🧍🧍🧍🧍🧍
Chicken	🧍🧍🧍🧍🧍🧍🧍🧍🧍🧍
Pasta	🧍🧍🧍🧍🧍🧍🧍
P.B. & J.	🧍🧍🧍🧍🧍🧍🧍🧍

Key: Each 🧍 equals 5 students

Outcome Table

Spinner 1

Color	Prediction	Outcome

Spinner 2

Color	Prediction	Outcome

Spinner 3

Color	Prediction	Outcome

Spinner 1

Color	Prediction	Outcome

Spinner 2

Color	Prediction	Outcome

Spinner 3

Color	Prediction	Outcome

Spinner 1

Color	Prediction	Outcome

Spinner 2

Color	Prediction	Outcome

Spinner 3

Color	Prediction	Outcome

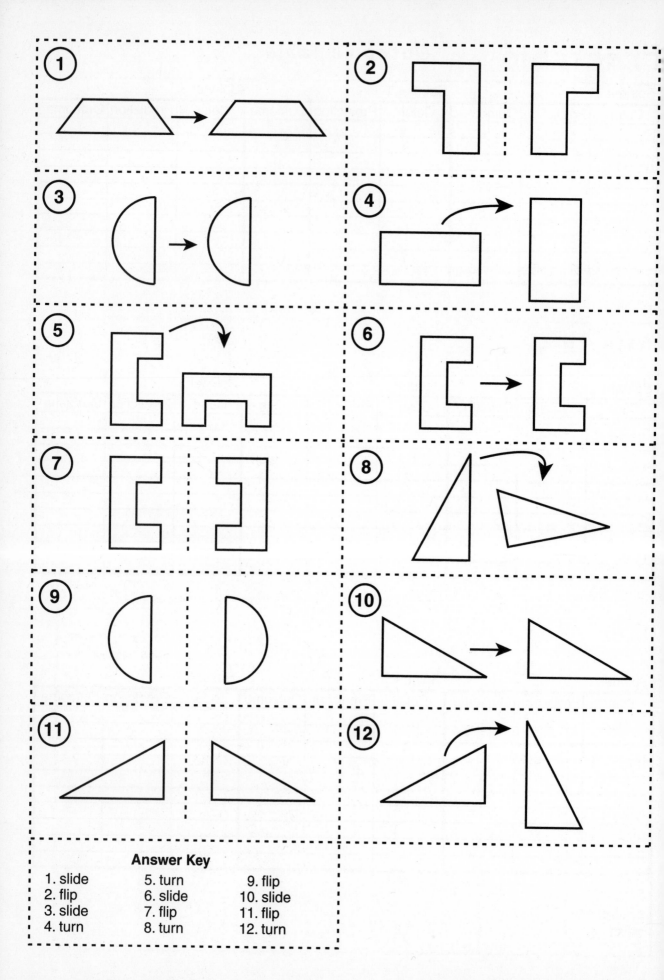

Answer Key

1. slide	5. turn	9. flip
2. flip	6. slide	10. slide
3. slide	7. flip	11. flip
4. turn	8. turn	12. turn

Fractable

Model						
Number of parts	2	3	4	5	6	7
Number of shaded parts						
Fraction of shaded parts						

Fractable

Model						
Number of parts	2	3	4	5	6	7
Number of shaded parts						
Fraction of shaded parts						

0.44	0.21
1.81	2.25
2.1	3.15
0.19	0.9
1.36	1.55
2.6	2.66
3.78	3.95
0.84	0.62
1.23	2.34
3.45	2.99

How would you measure each item?
Use centimeter (cm), decimeter (dm), or meter (m).

| | 1. | Measure length of swimming pool. |

1. Measure length of swimming pool.

2. Measure length of referee's whistle.

3. Measure length of track for running a race.

4. Measure length of tennis racket.

5. Measure width of baseball diamond.

6. Measure width of bicycle tire.

7. Measure length of arrows for archery.

8. Measure size of table tennis balls.

9. Measure length of soccer field.

10. Measure width of wheels on roller skates.

11. Measure height of basketball goal.

12. Measure length of a paper clip.

Answer Key:

1.	m	7.	dm
2.	cm	8.	cm
3.	m	9.	m
4.	dm	10.	cm
5.	m	11.	m
6.	cm or dm	12.	cm

Garden Gate

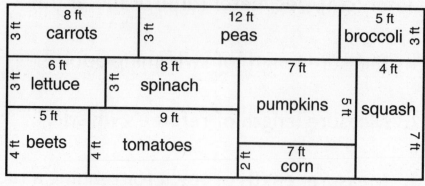

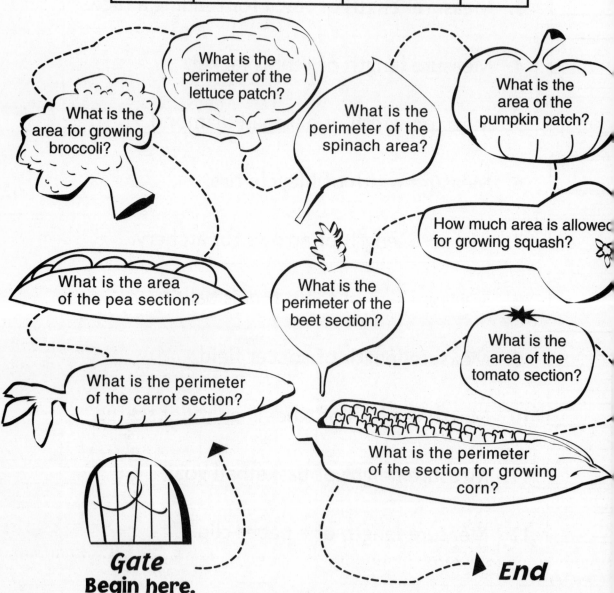

What is the area for growing broccoli?

What is the perimeter of the lettuce patch?

What is the perimeter of the spinach area?

What is the area of the pumpkin patch?

How much area is allowed for growing squash?

What is the area of the pea section?

What is the perimeter of the beet section?

What is the area of the tomato section?

What is the perimeter of the carrot section?

What is the perimeter of the section for growing corn?

Gate
Begin here.

End

Answer Key
(Perimeter) Carrots = 22 ft
(Area) Peas = 36 sq ft
(Area) Broccoli = 15 sq ft

(Perimeter) Lettuce = 18 ft
(Perimeter) Spinach = 22 ft
(Area) Pumpkins = 35 sq ft
(Area) Squash = 28 sq ft

(Perimeter) Beets = 18 ft
(Area) Tomatoes = 36 sq ft
(Perimeter) Corn = 18 ft

Division Toss		
Number	**Divide by**	**Points**
	2	
	3	
	4	
	Lucky Space Points = Number Tossed	
	5	
	6	
	7	
	Lucky Space Points = Double the Number Tossed	
	8	
	9	
	Total	

Remainders:

Division Toss		
Number	**Divide by**	**Points**
	2	
	3	
	4	
	Lucky Space Points = Number Tossed	
	5	
	6	
	7	
	Lucky Space Points = Double the Number Tossed	
	8	
	9	
	Total	

Remainders:

Division Toss		
Number	**Divide by**	**Points**
	2	
	3	
	4	
	Lucky Space Points = Number Tossed	
	5	
	6	
	7	
	Lucky Space Points = Double the Number Tossed	
	8	
	9	
	Total	

Remainders:

Division Toss		
Number	**Divide by**	**Points**
	2	
	3	
	4	
	Lucky Space Points = Number Tossed	
	5	
	6	
	7	
	Lucky Space Points = Double the Number Tossed	
	8	
	9	
	Total	

Remainders:

For use with Chapter 28 Practice Game

Start

Finish

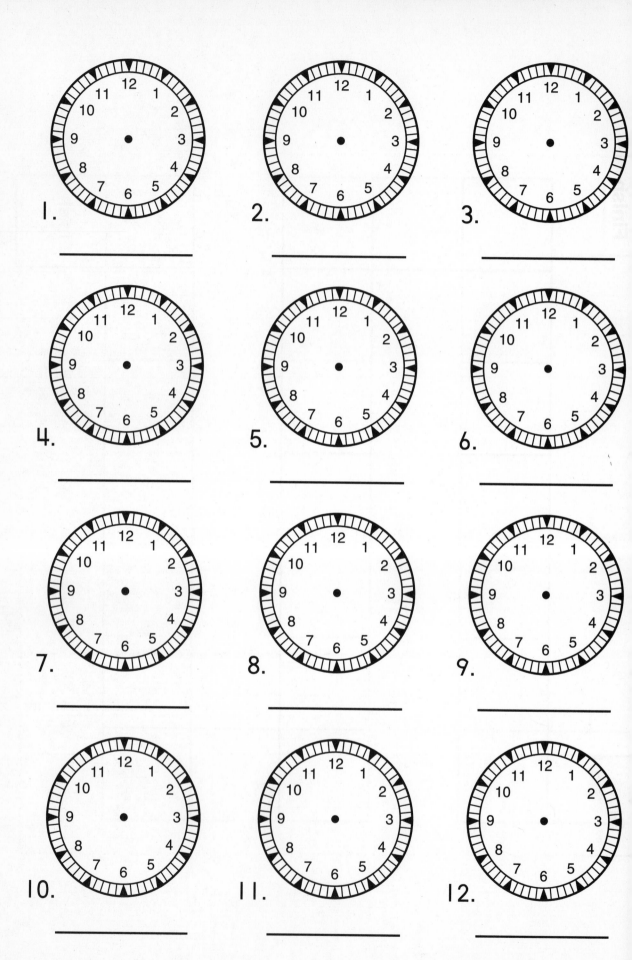

1. _____

2. _____

3. _____

4. _____

5. _____

6. _____

7. _____

8. _____

9. _____

10. _____

11. _____

12. _____

Start

| 45 ÷ 5 | 3 × 7 | 2 × 9 | 16 ÷ 4 | 8 × 2 | 15 ÷ 3 |

6 × 9

| 4 ÷ 4 | 4 × 6 | 18 ÷ 2 | 9 × 7 | 12 ÷ 3 |

9 × 3

| 18 ÷ 3 | 5 × 7 | 32 ÷ 4 | 27 ÷ 3 | 8 × 8 |

30 ÷ 5

Finish

| 24 ÷ 3 | 6 × 3 | 20 ÷ 4 | 7 × 5 | 21 ÷ 3 | 4 × 5 |

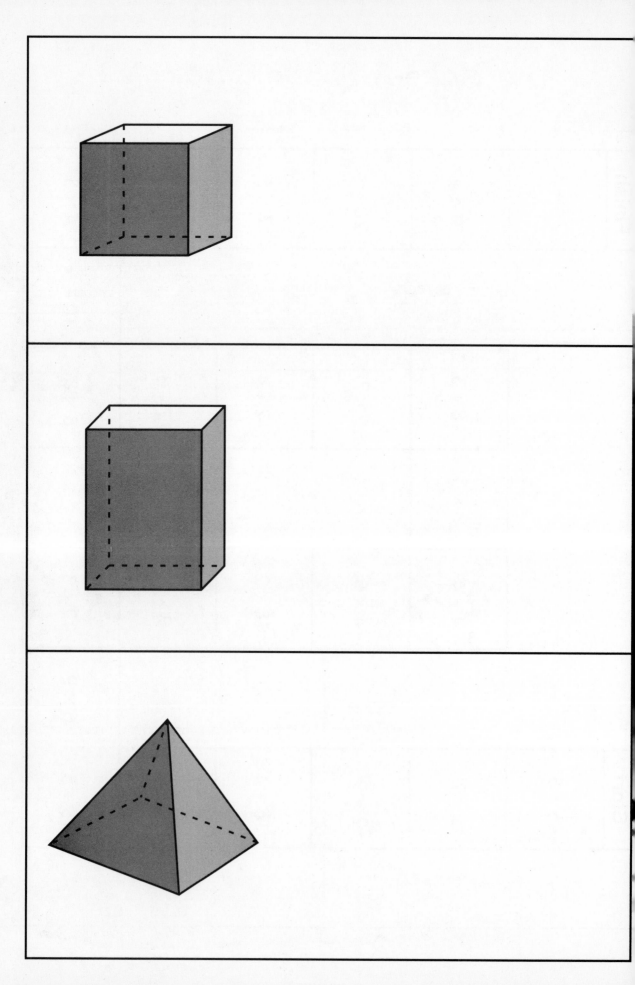

Make-A-Shape Cards

Mark the following points:
(1, 1), (3, 1), (1, 3) and (3, 3).

Mark the following points:
(0, 5), (0, 6), (3, 5) and (3, 6).

Mark the following points:
(4, 7), (5, 8) and (6, 7).

Mark the following points:
(0, 0), (1, 1), (0,1) and (1, 0).

Mark the following points:
(0, 7), (2, 7), (0, 8) and (2, 8).

Mark the following points:
(4, 1), (4, 4), (5, 4), (6,1) and (6, 3).

Mark the following points:
(2, 6), (4, 6) and (3, 8).

Mark the following points:
(1, 4), (1, 5), (4, 4) and (4, 5).

$\dfrac{1}{2}$	$\dfrac{1}{3}$	$\dfrac{2}{3}$
$\dfrac{1}{4}$	$\dfrac{2}{4}$	$\dfrac{3}{4}$
$\dfrac{1}{6}$	$\dfrac{2}{6}$	$\dfrac{3}{6}$
$\dfrac{4}{6}$	$\dfrac{6}{6}$	$\dfrac{1}{8}$
$\dfrac{2}{8}$	$\dfrac{3}{8}$	$\dfrac{4}{8}$
$\dfrac{5}{8}$	$\dfrac{6}{8}$	$\dfrac{8}{8}$

	Lengths	
What I measured	**Estimate**	**Measurement**
_____	about _____ in.	about _____ in.
_____	about _____ in.	about _____ in.
_____	about _____ in.	about _____ in.
_____	about _____ in.	about _____ in.
_____	about _____ in.	about _____ in.
_____	about _____ in.	about _____ in.
_____	about _____ in.	about _____ in.
_____	about _____ in.	about _____ in.
_____	about _____ in.	about _____ in.
_____	about _____ in.	about _____ in.
_____	about _____ in.	about _____ in.

Find Me Cards

Find an object 1 cm in length.	Find an object 5 cm in length.	Find an object 10 cm in length.
Find an object 15 cm in length.	Find an object 20 cm in length.	Find an object 50 cm in length.
Find an object 1 m in length	Find an object 2 m in length	Find an object or part of the classroom that is 5 m in length.